Canines in the Wild

WOLVES IN THE WILD

Natalie Humphrey

Gareth Stevens PUBLISHING

Please visit our website, www.garethstevens.com. For a free color catalog of all our high-quality books, call toll free 1-800-542-2595 or fax 1-877-542-2596.

Cataloging-in-Publication Data
Names: Humphrey, Natalie.
Title: Wolves in the wild / Natalie Humphrey.
Description: New York : Gareth Stevens Publishing, 2023. | Series: Canines in the wild | Includes glossary and index.
Identifiers: ISBN 9781538279632 (pbk.) | ISBN 9781538279656 (library bound) | ISBN 9781538279649 (6pack) | ISBN 9781538279663 (ebook)
Subjects: LCSH: Wolves–Juvenile literature.
Classification: LCC QL737.C22 H86 2023 | DDC 599.773–dc23

Portions of this work were originally authored by Holden Strauss and published as *Wolves*. All new material this edition authored by Natalie Humphrey.

Published in 2023 by
Gareth Stevens Publishing
29 East 21st Street
New York, NY 10010

Editor: Natalie Humphrey
Designer: Rachel Rising

Photo credits: Cover Pat Cacioppo/Shutterstock.com; cover, p. 1 Creative Travel Projects/Shutterstock.com; cover, pp. 1, 3, 4, 6, 8, 10, 12, 14, 16, 18, 20, 22–24 Attitude/Shutterstock.com; cover, pp. 1, 3, 4, 6, 8, 10, 12, 14, 16, 18, 20, 22–24 donatas1205/Shutterstock.com; cover, pp. 1, 3, 6, 8, 10, 14, 20, 22–24 Olga_C/Shutterstock.com; p. 4 belizar/Shutterstock.com; pp. 5, 7 Volodymyr Burdiak/Shutterstock.com; p. 9 Ondrej Prosicky/Shutterstock.com; p. 11 Jim Cumming/Shutterstock.com; p. 12 Olly S. Cannon/Shutterstock.com; p. 13 Grindstone Media Group/Shutterstock.com; p. 15 Bildagentur Zoonar GmbH/Shutterstock.com; p. 16 Stepan Kapl/Shutterstock.com; p. 17 PNQ Images/Shutterstock.com; p. 18 photomaster/Shutterstock.com; p. 19 Michal Martinek/Shutterstock.com; p. 21 worldswildlifewonders/Shutterstock.com.

Printed in the United States of America

CPSIA compliance information: Batch #CSGS23: For further information, contact Gareth Stevens, New York, New York, at 1-800-542-2595.

CONTENTS

Words in the glossary appear in **bold** type the first time they are used in the text.

A HOWL IN THE NIGHT

With a **howl** that can be heard from over 10 miles (16 km) away, wolves are the subject of many **legends** around the world. Found in North America, Europe, and Asia, the wolf is the largest member of the **canine** family.

Despite their legendary image, wolves aren't as dangerous as they seem. This wild canine would rather stay as far away from humans as it can. Wolves rarely attack humans. It's usually humans that hunt them!

The gray wolf, also called the timber wolf, is the heavist wolf species.

NORTH AMERICAN WOLVES

Throughout the world, there are many different types of wolves. In North America, scientists agree there are at least three different **species** of wolf. These species are the gray wolf, the red wolf, and the Arctic wolf.

The largest in size of the three species is the gray wolf. Gray wolves can also be found on other continents. The red wolf is smaller and can now only be found in North Carolina. The Arctic wolf lives in northern Canada and Greenland.

That's Wild!

Scientists believe that **domestic** dogs and many other wild canines came from early gray wolves. These early wolves were different from the gray wolves today.

Wolf paws can be over 4 inches (10 cm) wide and 4.5 inches (11 cm) long!

WHAT DOES A WOLF LOOK LIKE?

Wolves come in many different colors. A wolf's fur can be gray, tan, brown, black, reddish, or white. The color of a wolf's fur often depends on where it lives. Gray wolves blend into rocks and trees, while white Arctic wolves blend in with snow.

Wolves also come in different sizes. Wolves that live in hot, dry areas are lighter in weight and color. The Ethiopian wolf resembles a red fox in color and weighs between 24 and 42 pounds (11 and 19 kg). Gray wolves that live in cold regions can be up to 143 pounds (65 kg)!

That's Wild!

Wolves are double coated, which means their coats have two layers instead of just one. These special coats keep wolves warm in the winter, but also cool in the summer!

The Ethiopian wolf, only found in Africa, is one of the **rarest** wolves in the world.

A WOLF PACK

Wolves are very social animals and often live in a close family unit called a pack. Packs are often made up of two wolf parents, also called a breeding pair, and their pups, or baby wolves.

Pack life is very important for wolves. Wolf packs keep and protect a **territory** together as well as hunt together. Depending on the location and the species, wolf packs can be different sizes. A typical wolf pack is around six wolves, but some wolf packs have more than 20 members.

That's Wild!

Bigger isn't always better for wolves. Larger wolf packs can take down bigger **prey**, but more wolves means there are more mouths to feed!

Pack sizes regularly change depending on the species, time of year, and amount of food available.

ADAPTED FOR THE WILD

Wolves have perfectly **adapted** to their home. Their pointed ears can move in different directions at the same time. Under the right conditions, a wolf can hear sounds from more than 6 miles (9.6 km) away and pick up higher **pitch** sounds than humans can.

When looking for prey, a wolf can pick up a scent about 1.7 miles (2.7 km) away. When they catch the scent of prey, wolves run fast after it. Wolves don't tire easily and can travel more than 30 miles (48 km) in one day!

At top speed, an Arctic wolf can run over 46 miles (74 km) per hour.

A BIG FAMILY

During the spring, female wolves will give birth to a **litter** of pups. Usually, there are between four and six pups per litter. The entire pack takes care of the young pups. When they are old enough, the pups will join the hunt with the other wolves.

When wolves are between one and two years old, they will often leave the pack to start their own. Young wolves will howl to let other lone wolves know where they are. When they find a match, wolves start their own pack.

That's Wild!

Roughly three weeks before her pups are born, female wolves will dig out a den. After the pups are born, the female wolf will usually stay in her den with her pups until they're old enough to leave on their own.

Wolf pups are usually 1 pound (0.5 kg) at birth.

WHAT'S FOR DINNER?

Wolves are carnivores, which means they eat mostly meat. On their own, wolves will hunt smaller prey like birds, fish, snakes, or rabbits. However, when members of a wolf pack come together to hunt, they look for much larger prey.

A successful hunt for a wolf pack might land them a deer or elk. They might also catch a caribou or moose. Other large meals might include huge bison or musk oxen. A gray wolf can eat up to 22.5 pounds (10.2 kg) of meat at one time!

If they need to, adult wolves can survive for days without food.

MASTER HUNTERS

When it's time to catch a meal, wolf packs work together to take down their prey. The pack **cooperates** to plan the attack.

When a wolf pack spots a herd of prey animals, they'll follow the herd for days. They watch the herd and find the weakest member. When the time is right, several members of the pack will run after the herd at once. They'll chase the herd until the weakest member gets separated from the herd. Then, the pack takes the prey down together.

Wolves usually live up to 12 years in the wild.

LIFE ALONGSIDE WOLVES

In 1926, the last native pack of gray wolves in Yellowstone National Park was killed. People thought that getting rid of wolves would help the **ecosystem**, but it did the exact opposite.

Without the gray wolves, elk populations grew to the point where they started destroying natural plant life. Without the plants, other species couldn't survive. When gray wolves were brought back into Yellowstone, they brought balance with them. Wolves help keep prey populations under control, which helps all species in their ecosystem.

That's Wild!

In 1995, scientists brought 14 wolves from Jasper National Park in Alberta, Canada, to Yellowstone. These wolves started families, and by 2020, 94 wolves and eight packs were recorded in Yellowstone!

Currently, the largest wolf pack recorded was over 37 wolves.

GLOSSARY

adapt: To change in order to live better in a certain environment.

canine: Of or relating to dogs.

cooperate: To work together toward the same goal.

domestic: Bred and raised for use by people.

ecosystem: A natural community of life that includes living and nonliving features.

howl: A long, loud cry that sounds sad or scary.

legend: A story that has been passed down for many years that's unlikely to be true.

litter: A group of pups born at the same time.

pitch: How low or high a sound is.

prey: An animal that is hunted by other animals for food.

rare: Uncommon or special.

species: A group of plants or animals that are all of the same kind.

territory: An area of land that an animal considers to be its own and will fight to defend.

FOR MORE INFORMATION

BOOKS

Ardely, Anthony. *What If Gray Wolves Disappeared?* New York, NY: Gareth Stevens Publishing, 2023.

Storm, Marysa. *Gray Wolves.* Mankato, MN: Black Rabbit Books, 2022

WEBSITES

National Geographic Kids: Gray Wolf
kids.nationalgeographic.com/animals/mammals/facts/gray-wolf
Learn more about gray wolves and where they live.

San Diego Zoo: Wolf
animals.sandiegozoo.org/animals/wolf
Check out more interesting facts about wolves in the wild.

INDEX